I0485162

WHAT'S INSIDE:

Contents

WITH APPRECIATION:

How does inspiration arise?   Sometimes it's in the middle of the night and once in the shower!   This time it is clear: Bill Woodward suggested there were possible contamination points at test equipment. Bill Woodward's classic ETA study guide, *"Fiber Optic Installer"* trains for the future and re-affirms important tenets of these works in scientific terms.

As described in "How To Precision Clean a Fiber Optic Connection", there are myriad physical venues and ambient contamination sources. In this instance, possible contamination is at the actual test equipment "port" that receives the connection. An early study in 2003 resulted in a 'ferrule cleaning swab tool". It is only with recent advances in speed and capacity that we now must consider all potential "weak links" to assure success for the magnificent medium that is a fiber optic transmission.

No works of this detail could be accomplished without many inputs. These surely came from individuals such as Eric Martini and Paul Blair at ITW Chemtronics. Eric's understanding of 'solvent cleaning'; Paul's knowledge of 'wiping materials' provided critical inputs during my time there.

Another early influence is Glenn Porter of Microenterprises and he remains so to this day. His study of fiber optic microscopy and efforts to create an automatic, computer-assisted inspection system deeply influenced the thesis of Five Zone Inspection and matters such as three-dimensional aspects of dry debris, fluidic contamination and combinations of the same. For many years I studied with his ME-9600. (now out of production) Some of Glenn's vintage equipment is featured here.

Mike Schneider, founder of Optical Design Manufacturing (ODM) carries forward some of Porter's work with his own innovation: a range of portable instruments with wide "field of view" and astoundingly brilliant resolution that (at times) seems to show images like an interferometer. Mike's latest VIS-300 video scope is featured herein. It is a trusted resource.

There are others: Jim Hayes at The Fiber Optic Association and Larry Johnson, Founder of The Light Brigade have been not only encouraging but highly influential for me. Larry and Jim have contributed immensely to our Industry. As has Frank Giotto of Fiber Instrument Sales: his team that include Kim Teesdale and Kirk Donley have impacted the market and influenced me for the better. John Cotterill at JSC Aeroptics always offered the best advice with an amazing sense of humour! Thank you, Ron Vigil and Mike Farrell. *And surely, there are dozens other who cannot be named for reasons those in the Industry understand very well.*

In recent times I have been honored to associate with former "competitors and foes"! Among these are Brian Teague at MicroCare who has provided important insights to these important topics.

Finally, in the first place, I want to thank my wife, Lanet, and others in the family who endured the times during all hours of the night, perhaps on vacation, I would be tapping keys on the computer as they calmly closed the bedroom door in places all over this wonderful world!

Thank you, all.

February-2015

2

INTRODUCTION:

Until recently, the fiber optic connection and possible transmission limiting dry debris, fluidic contamination and combinations of both have been categorized in two dimensions.

Standards based on IEC 61300-3-35 and IEC TR-62627 as well as those such as TIA 455-250 and Telcordia GR-2923-Core refer to the end face only in the "horizontal plane" when the reality is a "vertical plane" as well as other associated three dimensional "geometry".

Likewise, debris and contamination are standardized in micron 2D range of diameter. Interferometer readings clearly show the 3D height of dry debris and fluidic contamination that can exceed diameter.

As a carry on work from *"How To Precision Clean All Fiber Optic Connections"* and "A *Study of Precision Cleaning Processes: What Works and What Does Not"*, this tutorial explains how commonly used equipment can be a "contamination source".

Since there is no practical way to inspect the test or inspection equipment, a preventative maintenance program is suggested.

This work suggests how contamination points on test and measurement equipment can influence readings as easily as touching an end face with oily fingers or dropping the jumper into a sandy soil.

- A dark smudge
  on a piece of white paper …

- A finger print on clean glass…

"CROSS-CON·TAM·I·NA·TION" *

NOUN: CROSS-CONTAMINATION
THE PROCESS BY DUST. FLUIDS OR OTHER
MICRO PARTICLES ARE UNINTENTIONALLY TRANSFERRED
FROM ONE SUBSTANCE OR OBJECT TO ANOTHER,
WITH HARMFUL EFFECT.

THE <u>TRANSFER</u> OF A <u>CONTAMINANT</u> FROM ONE SOURCE TO ANOTHER.

* February 1st, 2015 Google Search in Wictionary™

- There are myriad examples. In the instance of precision cleaning a fiber optic connection not only can the connection end face and associated geometry be contaminated by environmental sources, *<u>test equipment itself can be a contamination source.</u>*

- Proper maintenance of anything that has a "test port": video inspection, OTDR, light source, and, *all instruments that may be in contact with a fiber optic end face connection* must be scrutinized and properly cleaned to assure accurate readings and long life of these important investments.

- Far too many connections are cleaned without video inspection: Please, when you are "given" one of these instruments, extracted from a limited budget, maintain it.

What is Being Cross-Contaminated: There are three general types of fiber optic inspection.

Type 1: Portable, direct view microscope. Not unlike common binoculars or a lab microscope, this instrument is never recommended to view an active laser.

Type 2: Portable "true" video inspection scope that uses a camera to project the end face image.

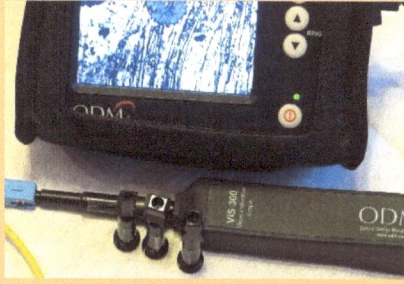

Type 3: Laboratory video inspection has a camera. Typically not suitable for field user.

Did You Know?

Video Inspection is the only way to determine if the connection is actually clean.

What does Type-1, Type-2 and Type-3 Fiber Optic microscopy have in common?

Adapters and Scrollers

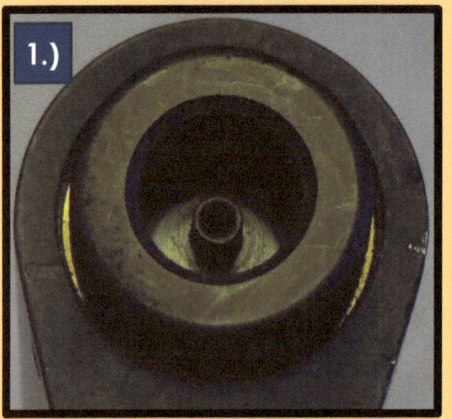

1.)

2.)

1.) The "port" that accepts the adapter type that connects the fiber optic connection to the inspection instrument

2.) A viewing lens or screen

3.) Typically, the Type-1 device only views one size fiber connector.

4.) Type 2 and 3 devices exchange adapters meaning they can be used on multiple connector types

WHAT'S A 'SCROLLER'?

Most fiber optic inspection only sees a specific area of the end face. Some instruments are "self-centering" and others are not.

The "scroller" enables a scientist to move the field of view to see sections of the connection that are not seen by typical inspection devices.

WHAT ARE THE CROSS-CONTAMINATION POINTS ON THE TYPE-1 FIBER OPTIC INSPECTION INSTRUMENT?

WHY NOT COTTON?

Cotton is an ideal material for "thread" and here you see cotton "threading". (1) This lint material can cross contaminate the port or be transferred to the end face.

WHY NOT PAPER?

Paper is absorbent, but it does not have tensile strength. (2) When it tears it shreds. When it absorbs too much liquid, it disintegrates. (3)

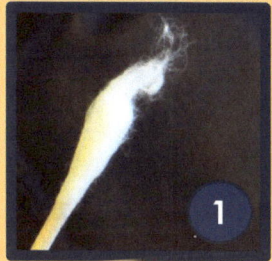
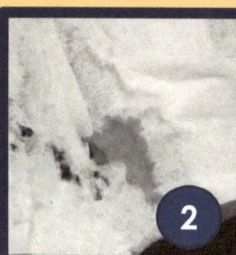
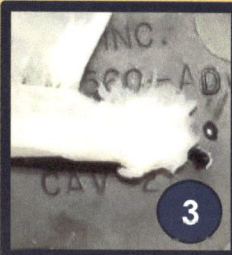

WHAT ARE THE CROSS-CONTAMINATION POINTS ON THE TYPE-1 FIBER OPTIC INSPECTION INSTRUMENT?

HOW IS THE INSTRUMENT CLEANED?

a. Select a fiber optic grade swab tool to match the diameter of the "port".

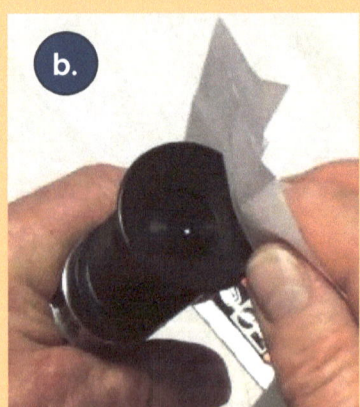

b. Select a lens-grade optical tissue. These are available from many sources.

Never use a "lens-grade tissue", dry or pre-moistened to precision clean the fiber optic end face. WHY? These tissues may contain soaps, ESD (electrostatic discharge) treatments and other components not acceptable for end face cleaning.

8

WHAT ARE THE CROSS-CONTAMINATION POINTS ON THE TYPE-2 FIBER OPTIC INSPECTION INSTRUMENT?

HOW IS THE INSTRUMENT CLEANED?

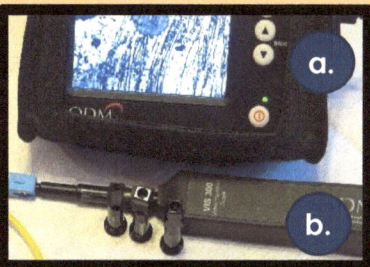

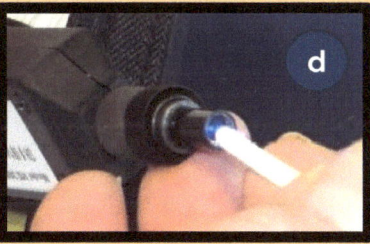

The Type-2 Portable Video Inspection Scope has two main components: a.) The Monitor and b.) The probe with tip assortment.

On the tip of the probe is a removable adapter. c.) Adapters match the size and configuration of the end face or back plane. and assure accurate inspection. Adapters screw on or twist-lock.

d.) The adapter lens is cleaned with an appropriate clean room grade swab tool.

What are the cross-contamination points on the Type-1 Fiber Optic Inspection Instrument?

How is the instrument cleaned?

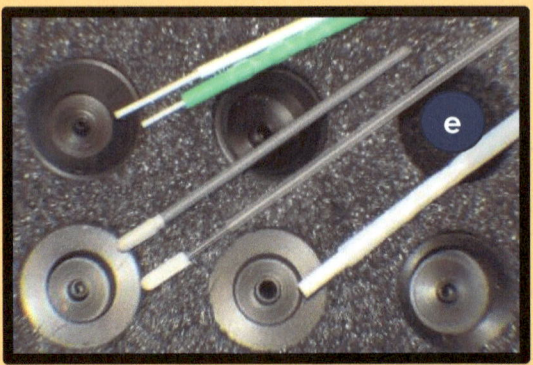

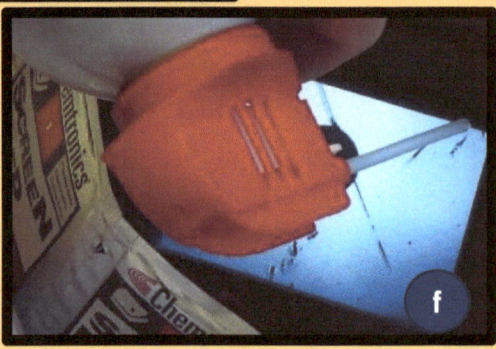

e.) Select a swab tool to match the configuration of the adapter. Moisten it.

f.) Clean the monitor screen with pre-moistened lens grade tissues. Using a moist wiper works best to capture dust and grime. *Do not use for end face cleaning.*
Dust can be moved with an "upside-down valve" compressed gas duster.

___

Caution: Never use a compressed gas duster to clean an end face. Why: 1.) It does not work, 2.) Damage potential to small fiber conductors.

THE "LIGHT" SEEN IN A VIDEO SCOPE IS NOT
TYPICALLY A LASER. IT IS A VIDEO CAMERA.
ALWAYS READ THE MANUFACTURER'S
SPECIFICATIONS. AN ACTIVE LASER CAN
DAMAGE EYE SIGHT AND VIDEO INSPECTION IS
BOTH A PERFORMANCE AND SAFETY ADVANCE.

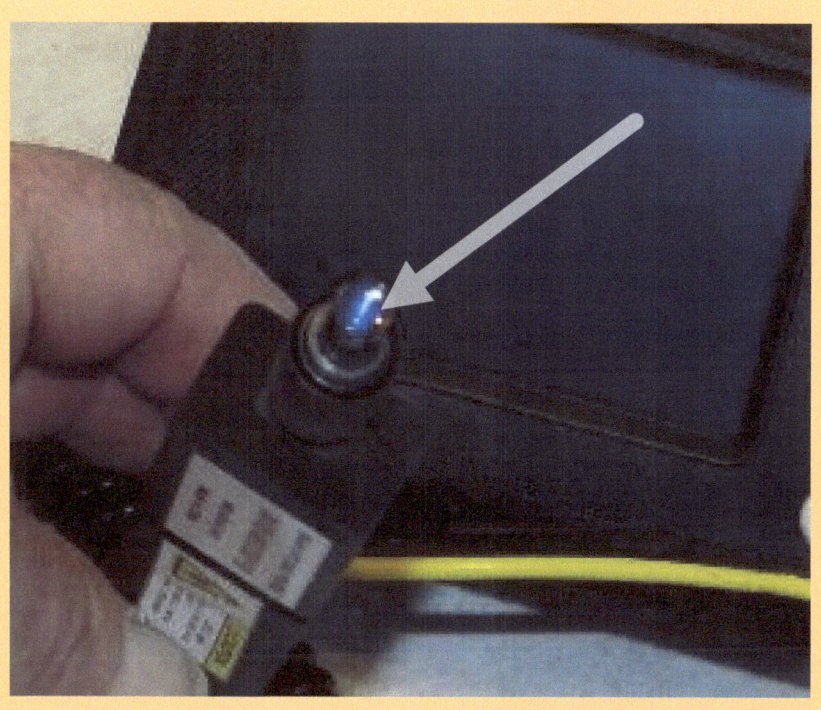

WHAT ARE THE CROSS-CONTAMINATION POINTS ON THE TYPE-3 FIBER OPTIC INSPECTION INSTRUMENT?

HOW IS THE INSTRUMENT CLEANED?

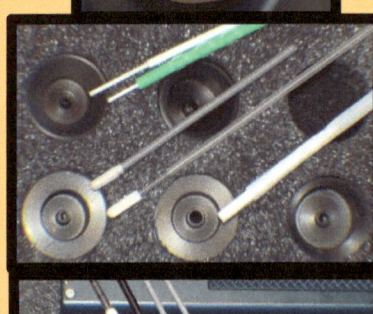

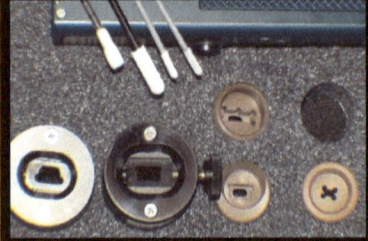

The Type-3 Laboratory Video Inspection Scope is likely to have more components…if only in the adapter selection.

As we studied in *"How To Precision Clean All Fiber Optics"* and the comparison/companion work, "*A Study Of Precision Cleaning Methods for all Fiber Optics*", we learned that debris and contamination are attracted to moisture.

Dry Cleaning works to "mop" fluidic contamination but is not effective for dust removal. <u>WHY?</u> There is an electrostatic discharge phenomenon called a tribo charge. This is a static field that can be established that attracts more dry soil as an unwilling part of the clean procedure.

HOW TO MOISTEN A SWAB TOOL: touch the tip in a small spot of precision cleaner for a count of 1-2-3-4-5. Insert the swab into the adapter. Some swabs pass-through the adapter which further decreases cross-contamination. This is an excellent way to clean not only video inspection, but also OTDR, and all "ports" commonly used in telecommunications.

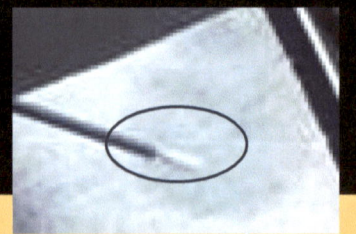

What are the cross-contamination points on the Type-1 Fiber Optic Inspection Instrument?

How is the instrument cleaned?

Beneath the adapter on the Type-3 Laboratory Video Inspection Scope is a precision lens. Select a clean room grade swab tool. These are polyester and will not leave a residue while absorbing minute dust particulates. You will be able to "see" the lens to assure it's clean...much like looking at your eyeglasses!

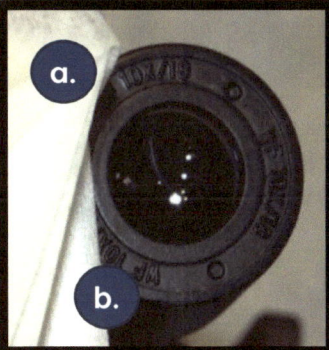

The eye piece is much like the lens on a microscope or binoculars. It should be cleaned with a precision wiper. There are two shown here: a.) is a lens grade tissue and b.) is a 'hydroentangled' polyester/cellulose blend. Paper wipers have their place...it is not in precision cleaning.

13

How do you know it's clean?

Of course, there is no practical way to inspect the inspection other than in these two ways:

1.) Observe the lens or monitor. Is the surface still streaked or dirty?

This debris was removed from the exterior of the probe... it could have cross contaminated.

2.) Look at the cleaning material. *Have you ever wondered why high-quality swab tools and wipers are white?*

It's because you will be able to see debris on the white background.

Always look at the wiper or swab to assure you have removed the soil!

Simple observations really work!

THE FIBER OPTIC END FACE, DRY DEBRIS AND FLUIDIC CONTAMINATION ARE 3D STRUCTURES!

Existing standards in 2015 only consider an area of a radius of the core in a Zone System such as: Zone 1-2-3 and both this area and contamination in 2D.

The topic is not complicated other than by standards such as IEC61300-3-35 and IEC TR-62627 which have fostered TIA, ARINC, IEEE and others. These standards consider the fiber optic end face in a limited area and base cleaning procedures on contamination that is easy to remove.

In reality, the connection and contamination (as all things) are three dimensional. It is from this 3D perspective cross-contamination can begin from "Zone-5" of a ferrule, for example, and transfer through the adapter to the instrument. This can skew the result recorded by a technician.

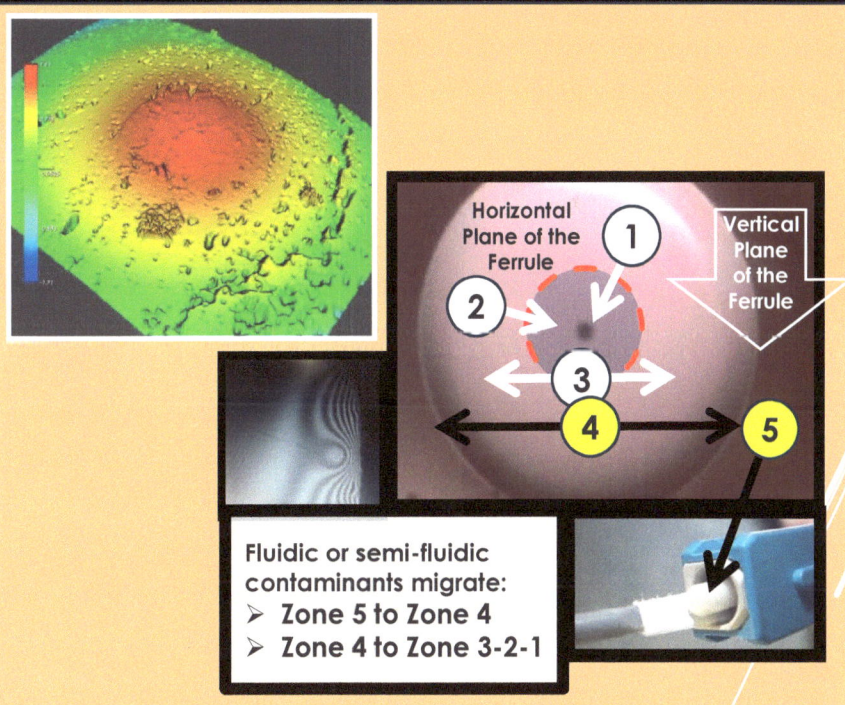

Horizontal Plane of the Ferrule

Vertical Plane of the Ferrule

1
2
3
4
5

Fluidic or semi-fluidic contaminants migrate:
➢ Zone 5 to Zone 4
➢ Zone 4 to Zone 3-2-1

How to select a fiber optic cleaner:

Worker safety and environmental concerns are always top criteria. Be sure you read and have MSDS explained to you: ask the provider. Merely *having* an MSDS is not the point OSHA intended to educate you...it's important to understand the document itself.

This range of chemicals represents the 2015 list of best choices. These are general chemical families and each company has a 'trade name' for their product. Ask your supplier *"what's in this stuff?"*.

To "future-proof" you also have to self-educate.

HFE 7100 w/IPA/CZ®	Precision Hydrocarbon Formulations	Aqueous (Glycol Ethers)
▸ Advantages	▸ Advantages	▸ Advantages
▸ Numerous formulations	▸ Numerous formulations	▸ Newest formulations
▸ Very good cleaning	▸ Wide range cleaning	▸ Wide range cleaning
▸ Check	▸ Check	▸ Check
▸ Convenience containers	▸ Convenience Containers	▸ Growing demand in many segments
▸ Easy Ship	▸ Low cost	▸ Convenience Containers
▸ Aerosol	▸ Disadvantages	▸ Easy Ship
▸ Non-flammable	▸ DOT regulated shipping	▸ Lowest cost
▸ Disadvantages	▸ As with IPA	▸ Disadvantages
▸ Ultra-Fast Evaporating		▸ Must dry with a 'wet-to-dry" step
▸ Can leave residues		
▸ Highest cost		

Test and compare. Try not to let 'aroma' be a criteria unless there are specific and individual reactions. A chemical choice is based on many factors and the actual cleaning ability is the foremost consideration. Demand training from your supplier: it may be no charge or fee based.

SELECTING FIBER OPTIC CLEANERS

- **How well does the solvent clean?**
- **How is it packaged?**
- **Shipping important?**
- **What is the environmental impact?**

Selecting a fiber optic cleaner is not as straight forward as answering "yes" or "no" to these questions!

Perhaps the most important questions are: 1.) What is the debris and what product works best? 2.) The next one is: "what works best if I cannot inspect each connection".

One thing is clear: not all fiber optic cleaning solvents so the same job! Some work better than others and others 'not so much'. Some have environmental restrictions…and there are hefty fines in some places.

Cost of delivery is always an issue. However, it is important to understand that virtually all fiber optic cleaners can be shipped per DOT and IATA regulations.

Challenge your supplier by asking "good questions": don't accept a sales story!

17

How to select a fiber optic wiping material

Many types of cloth have been created since the beginning of time. Earliest came from animal skins which were superseded by weaving. Now, there are synthetic animal skins and woven cloth of natural and synthetic materials.

There is also a new-generation of material that weaves or hydro-entangles synthetic and natural materials. Selection of a strong material that does not shred or leave a surface residue is a critical concern to the precision fiber optic cleaning process.

100% Cellulose (paper)	Polyester / Microfiber	Hydroentangled polyester/cellulose

▸ Advantages
  ▸ Absorptive
  ▸ Convenience package
  ▸ Readily Available
  ▸ Lowest Cost
▸ Disadvantages
  ▸ Low sheer strength
    ▸ Tears, shreds with Lint residues
  ▸ Used for many applications
    ▸ Embedded in supply chain
	▸ Advantages
	  ▸ Absorptive on most debris
	  ▸ High sheer strength
	    ▸ Low Linting
	  ▸ Often used in probes and other cleaning devices
	    ▸ Readily available
	  ▸ Moderate cost
	▸ Disadvantages
	  ▸ Can create a static charge
		▸ Advantages
		  ▸ Highly absorptive
		  ▸ High sheer strength
		  ▸ Often used in cleaning platforms
		    ▸ Readily available
		  ▸ Moderate cost
		▸ Disadvantages
		  ▸ Many choices and not all perform to the same level.

Test and compare. 100% cellulose (paper) and 100% cotton are not recommended to precision clean a fiber optic connection.

Be cautious when using ESD topical cleaners: these are effective for lenses but *must never be used for the end face*. Cross-contamination can take many forms. Professional training to use professional products is 'important -stuff'.

SELECTION OF WIPING MATERIALS

Lens Grade Tissues and hydroentangled non-woven cellulose/polyester blends are some of the most cost effective and high performance wipers available.

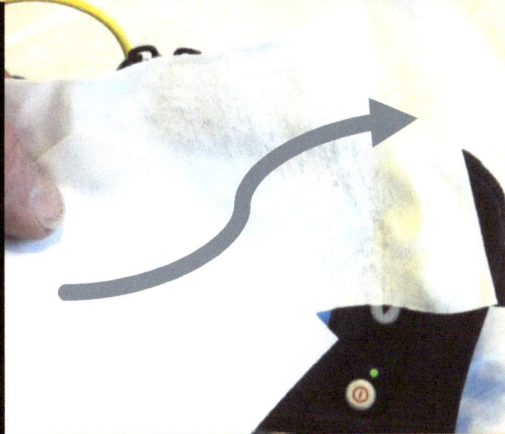

This 4"x4" cellulose/polyester wiper has exceptional strength. This is an indication it will not leave a lint residue. In this image, the wiper is distorted...but does not shred or tear.

However, there may be as many as 4,000 different cellulose/polyester combinations. Some have too much polyester and are not absorbent and others not enough and lack strength.

Challenge your supplier. Request samples.

Conclusions

The fiber optic medium is an ongoing deployment of new technologies.

- In 2015, transmissions in the gigabit ranges are common; in July-2015 a new world record of more than 40 terabits was set over a single fiber.

- This means that we, as researchers, designers and craftspersons must understand what some might called minor or nuanced matters: *in this case assuring all aspects of fiber optic transmission and test are calculated and studied.*

- While this work targets video inspection, also, other instruments such as OTDR, light source, power meters that may test fiber optic cables should be placed on a preventative maintenance schedule.

- What might that be? If the connection is housed in a clean environment, perhaps once a month. If the connection is made on a feed lot in Western Kansas, Military theatre in a desert region, or, entertainment center along a sandy beach...the PM might be daily. It is the environment and ever increasing demands and expectations that mandates increased awareness to assure the tools of the trade are up to the job.

- Assuring the equipment is clean helps assure long life of expensive investments.

Test Your Knowledge:

True	False	
		The Type-1 Inspection device uses a miniature camera
		The Type-2 Inspection device has a laser light source
		The Type-3 Inspection device uses a video camera
		It is safe to view all fibers through a Type-1 Inspection device
		The Type-1 device has interchangeable adapters
		Type 2 and Type-3 devices have interchangeable adapters
		Inspection is the only way to determine if a fiber optic connection is actually clean
		Fiber Optic grade wipers are high quality cellulose material
		Fiber Optic cleaners are 99.9% IPA because it is non-flammable
		There is no way to know if the port or adapter is actually clean

Answers: Test your knowledge

True	False	
	F	No camera: it's a microscope
	F	The Type-2 Inspection device a video camera; most typically no laser.
T		The Type-3 Inspection device uses a video camera
	F	It is not eye-safe to view all fibers through a Type-1 Inspection device
	F	Typically to use a Type-1 device for different connections requires a specific instrument.
T		Type 2 and Type-3 devices have interchangeable adapters
T		Inspection is the only way to determine if a fiber optic connection is actually clean
	F	Fiber Optic grade wipers are either cellulose/polyester hydroentangled or cleanroom microfiber material. "High quality cellulose material" does not exist!
	F	IPA is flammable. There are superior non-flammable and flammable fiber optic cleaners. Since the cleaner is used in minute quantities, flammability is only one criteria: performance is the key.
	F	Observation of the surface and discoloration of the white cleaning material is an excellent indication of cleanliness.

ABOUT THE AUTHOR:

Ed Forrest has been actively involved in specification and applications engineering of various precision cleaning applications for more than 25 years.   Previously employed at ITW Chemtronics® (retired July-2014) he was schooled to analyze precision and gross cleaning applications in a wide range of applications. In 2001 he began development of a program that resulted in formal approvals at all major telecommunications providers.

He has seven patents specifically in the areas or fiber optic precision cleaning with six products in production. He has marketing credits that include branding, training, and publication of materials. He innovated a chemical mid-span break-in for ribbon fiber. He has other patents pending.

He is active on fiber optic standards committees and is considered a SME in the study of fiber optic cleaning and inspection. His work is based on field experiences and the needs of designers, crafts persons and production line workers.

His practical thesis of "Five Zone Cleaning" is a look forward to the times when high speed and capacity of fiber optic transmission (even more) will be impacted by a contaminated or improperly cleaned connections. He has uniquely researched inspection of the 4th and 5th Zone and the influences of various debris and contamination as it is positioned on these areas of the connector.

He worked as an Electronics Manufacturer's Representative throughout the 1970's. He actively participated in the early introduction of some of the most fundamental electronic products in the changeover from analogue to solid state.   These included solid state components, consumer products including the first hand-held calculators, esoteric high fidelity, test equipment, games and other electronic products considered 'cornerstones' of the contemporary marketplace.
He has production credits in that Industry

He worked in a then-developing market segment in the Home Furnishings Industry. By coordinating North American and International Development, using an effective agency in Denmark he was able to work throughout Europe prior to the time of the EU. Incoordination with C.ITOH (est-1860) , he traveled and developed a Japanese market long before current interest in the important nations of The Pacific Rim. He initiated promotional activity in conjunction with USA Embassies, individual USA states resulting in active trade in Denmark, Sweden, Finland, Italy, Germany. Great Britain, nations in The Middle East and South Africa. He has production credits in that industry.

Early career as a Technical Representative, he resolved on-site consumer complaints in Union Carbide Corporation's Automotive Consumer Products group, career-forming experiences include introduction of Prestone® AntiFreeze as a Summer Coolant in a one year NASCAR race test and associated promotions.
He spent an innovative time as "rep" to Standard Oil of Ohio® as SOHIO® introduced "self-service fueling" to the market.   He competed in the market when brands like STP® and Wynn's® dominated consumer interest.

He is an active photographer, enjoys study of the ancients, and a hobbyist collector of esoteric high-fidelity. As a life-long SCCA member he competed "wheel-to-wheel" in more than 200 events at SCCA's National Championship level in cars he designed…'with a little help from his friends'. Married with a fascination for Weimaraners, he and his wife are often at the edge with three lovely specimens! They have travelled extensively through the USA and Europe.

23

Want to Learn More?

- Discussion of standards and why they are obsolete when they are published
- How to Create Internal Standards for your organization and why.
- "Science of Cleaning"
- How to clean a fiber optic connection when you don't have a video scope

68 Pages in full color. 6"x9" Paperback

www.createspace.com/ 5173068

www.amazon.com

How to Precision Clean All Fiber Optic Connections

- FACTORS THAT INFLUENCE CLEANING AND INSPECTION
- METHODS AND PROCEDURES
- SUGGESTIONS FOR NEW STANDARDS.

RMS (RaceMarketingServices™)
est·1974

Bringing Ideas Together™

Edward J. Forrest, Jr.   © 2015 All rights reserved
www.fiberopticprecisioncleaning.com
www.cedarkeyinstitute.com

A STUDY OF PRECISION CLEANING METHODS FOR ALL FIBER OPTIC CONNECTIONS

- COMPARATIVE EVALUATIONS OF CONTEMPORARY PRODUCTS, METHODS AND PROCEDURES
- A STUDY OF THE SFP-TYPE OPTICAL TRANSCEIVER AND A 2.5MM JUMPER CONNECTION.
- WHAT WORKS; WHAT DOES NOT AND WHY.

RMS (RaceMarketingServices™)
est·1974

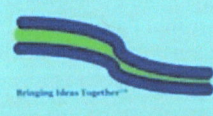

Bringing Ideas Together™

© Edward J. Forrest, Jr. 2015   All rights reserved

www.fiberopticprecisioncleaning.com
www.cedarkeyinstitute.com

- New Edition
- Answers the questions: what works best? CleTop®, IBC®, Swabs, QbE®, Sticklers®. FerruleMate®? ClePen.
- Discusses Zone-5 Contamination and how to avoid it.
- Evaluation and removal of "nasty soils" you might find on the job!

138 Pages in full color. 8.5x11" Paperback

www.createspace.com/ 5271116

www.amazon.com

Notes